Jailson Costa Gaspar
Gonçalo M. da Conceição
Regigláucia R. de Oliveira

Riacho do Ouro, Caxias, Maranhão, Brazil

Jailson Costa Gaspar
Gonçalo M. da Conceição
Regigláucia R. de Oliveira

Riacho do Ouro, Caxias, Maranhão, Brazil

Water Quality Parameters and Indicators

ScienciaScripts

Imprint

Cover image: www.ingimage.com

This book is a translation from the original published under ISBN 978-3-330-76196-4.

Publisher:
Sciencia Scripts
is a trademark of
Dodo Books Indian Ocean Ltd. and OmniScriptum S.R.L publishing group

120 High Road, East Finchley, London, N2 9ED, United Kingdom
Str. Armeneasca 28/1, office 1, Chisinau MD-2012, Republic of Moldova, Europe
Managing Directors: Ieva Konstantinova, Victoria Ursu
info@omniscriptum.com

Printed at: see last page
ISBN: 978-620-8-38339-8

The book Riacho do Ouro, Caxias, Maranhão, Brazil: Parameters of Water Quality, an initiative of the Laboratory of Plant Biology/LABIVE, Centre for Higher Studies in Caxias/CESC, State University of Maranhão/UEMA, makes available to the general public a work on aspects of water quality in Riacho do Ouro, a water body belonging to the Itapecuru River basin.

We congratulate the authors on the research carried out and the results achieved, as well as on their concern for the environmental problems of Maranhão's streams. We hope that other studies will be carried out, always with a view not only to raising problems, but also to pointing out solutions to environmental problems.

We wish a great read to all those who venture into acquiring new knowledge about environmental issues related to Maranhão's springs.

A big hug to you all!

Paula Regina Pereira Martins
PPGBAS/CESC/UEMA
Plant Biology Laboratory/LABIVE

SUMMARY

CHAPTER 1

INTRODUCTION

Biology teaches us that life cannot exist without water. All living beings need water for their survival, as it is part of the very constitution of living organisms and is also important in all physiological and metabolic processes, from the absorption of food to the elimination of waste. The importance of water is not only part of biology, but history reveals that the great civilisations arose in the valleys of great rivers, such as the Nile in Egypt, the Tigris-Euphrates in Mesopotamia, the Indus in Pakistan and the Yellow in India, with the implementation of large irrigation systems (BRUNI, 1994).

Water is the most abundant substance on the Earth's surface in terms of fresh and salt water. However, fresh water from surface runoff has a minimal fraction compared to salt water (WALDMAN, 2009). Of the total volume of water, 97% is marine and of the remaining 3%, 2% is available in rivers, lakes and groundwater (fresh water) and 1% is in hard-to-reach places such as snow, permanent glaciers and water vapours contained in the atmosphere (ROCHA; ROCHA, 2009).

Brazil has the largest water network with 17% of the planet's total volume, totalling around 6.2 billion m^3 of water. These figures were released by the United Nations Educational Scientific and Cultural Organisation (UNESCO) during the third edition of the World Water Forum in March 2003 in Kyoto, Japan (ROCHA; ROCHA, 2009).

According to Maranhão (2011) apud Leite (2011) 97.2 per cent of Maranhão's water resources are groundwater and 2.8 per cent surface water. These figures set Maranhão apart from other states. As well as being larger in volume than surface water, groundwater is consumed more by the population, using wells to tap into the subsoil.

Freshwater bodies are influenced by extrinsic environmental factors that interfere with quality indicator parameters such as pH, conductivity, oxygen, turbidity and others, making these bodies unstable. When it comes to aquatic environments, they are classified into two categories: lotic or flowing water habitats, which are represented by streams and rivers and are divided into rapids (or currents) and pools, while the other category is represented by lakes and ponds, both classified as lentic environments or stagnant water habitats (BROWN; LOMOLINO, 2006).

The National Environmental Council (CONAMA) Resolution No. 357 of 17 March 2005 classifies waters according to their salinity measured as a percentage, where watercourses with values <0.5% are considered fresh waters, for values >0.5 and 30% are classified as brackish waters and results 30% are considered saline waters (BRASIL, 2005).

Water is a non-renewable resource and because of this problem, there is increasing concern worldwide about the preservation of nature. In response to climate change in general, and the scarcity and quality of water, research is drawing attention to the rational use and preservation of water resources (DETONI; DONDONI; PADILHA, 2007).

In this context, some actions to control the rational use of water stand out, such as the decree of Law No. 9.433/97 known as the Water Law, drawn up through the National Water Resources Plan (PNRH), the systematic management of water resources, in relation to the quantity and quality of water, achieved through some objectives aimed at managing these resources made available to populations. In the future, humanity needs to establish certain standards for determining the quality of water according to its bodies of water and their use (BRASIL, 1997).

The State Water Resources Plan was consolidated through Decree Law No. 8.149/04 (MARANHÃO, 2004), establishing that responsibility for water resources lies with the State Secretariat for the Environment and Natural Resources (SEMA). This plan provides for programmes, standards, technical and administrative procedures and includes actions to ensure the quality and control of water use, as well as the implementation of preventive measures to reduce costs in the fight against pollution.

According to the classification of waters contained in current legislation, there are quality parameters to be met according to their classification. So if a given body of water has standards that are not in line with the predominant uses intended in its classification, it is necessary to establish mandatory measures and targets so that this watercourse once again reaches the quality standards of its appropriate classification (WEINBERG, 2013).

From this perspective, Barros et al. (2014) state that the municipality of Caxias in Maranhão lies on three hydrographic basins, the most important of which is the Itapecuru. The municipality has numerous watercourses, with the Itapecuru River having the largest volume and its tributaries, where several streams have been suffering serious environmental damage due to a lack of concern for water resources. There is visible contamination by sewage, deforestation of water sources, fires, silting up, extinction of flora and disregard for environmental protection laws.

CONAMA, through Resolution 357/2005, provides for the classification of water bodies and guidelines for their classification, as well as considerable limits that can be found through analysis and evaluation for each substance that determines water quality. It also describes water quality as the set of standards necessary to meet current and future predominant uses. Therefore, if a body of water manages to reach legal values outside the standard according to its classification, it could cause serious lethal damage or alterations in the behaviour of ecosystems (BRASIL, 2005).

Evaluating water quality through physico-chemical parameters and indicators is a way of identifying possible anthropogenic actions and natural alterations in the aquatic environment as a result of climate change or a way of establishing a quality index for watercourses according to their classification in current legislation (BRASIL, 2006).

According to Sánchez (2008), environmental degradation of anthropogenic origin causes alterations in the processes or functions that affect environmental quality, causing a negative impact on natural resources. From this perspective, environmental degradation in surface waters is caused by the dumping of urban, industrial and agricultural waste (VICTORINO, 2007). According to the author, this is a recurring problem in industrialised countries, which have strict legislation on water quality. It also works as a planning policy, helping to manage water resources as a thermometer aimed at monitoring the processes of water resources, characterising it in a qualitative way and targeting environmental control (GUEDES et al. 2012).

The research is justified by the fact that the Riacho do Ouro serves some of the communities around it and uses it for various domestic activities, including consumption. This research has therefore determined the water quality standards of the creek, which will contribute to improving the population's quality of life, as well as serving as guidelines for awareness-raising campaigns aimed at riverside populations, drawing attention to the conservation of the creek.

CHAPTER 2

LITERATURE REVIEW

2.1 Physical quality indicator parameters

2.1.1 Temperature

The temperature in water is represented by the kinetic energy of its molecules and synthesises the phenomenon that causes the transfer of heat to the environment. Temperature changes can be influenced naturally by the sun's rays or by industrial waste and water used to cool machinery (BRASIL, 2014). Temperature outside the thermal tolerance range can cause serious damage, affecting the growth and reproduction of aquatic organisms (CETESB, 2015). On average in Brazil, aquatic environments have a temperature of between 20°C and 30°C, with the exception of the southern region, which can reach 5°C at certain times of the year (BRASIL, 2014).

2.1.2 pH (Hydrogen Potential)

Hydrogen potential indicates the degree of acidity or alkalinity in a given medium. It is read using an antilogarithmic scale with a range from 0 to 14, indicating values below 7 for an acidic medium, above 7 for an alkaline medium and equal to 7 for neutrality (BRASIL, 2014).

The pH in natural waters can change naturally through rock dissolution, photosynthesis or through countless anthropogenic actions originating in domestic and industrial waste (BRASIL, 2014). These alterations can jeopardise the metabolism of aquatic organisms (CETESB, 2015) if they are outside the ideal range for maintaining life (6 to 9), although there are environments with values below 6 or above 9 caused naturally. In addition to damaging biota, pH changes can cause damage to installations through the appearance of corrosion and incrustation (BRASIL, 2014).

2.1.3 Conductivity

It is defined as the capacity of water to conduct electric current as a result of substances dissociated into anions and cations, expressed in mho (inverse of ohm) or S (Siemens) per unit of measurement of length cm (centimetres) or m (metre), the most commonly used unit being S. The higher the concentration of ions in natural waters, the greater the capacity of the water to conduct electric current, with values well above the range in waters polluted by domestic and industrial waste (BRASIL, 2014).

2.1.4 True colour

Physical quality parameter produced by dissolved particles with dimensions of less than 1 pm, with a reading compared to the sample with a cobalt-platinum standard, with the reading unit being uH (Hazen's unit). In order to characterise water, it is important to use the two parameters of apparent colour, which can be defined as dissolved and suspended particles, as opposed to true colour, which, through mechanical procedures (centrifugation or vacuum filtration), removes suspended materials from the reading (BRASIL, 2014).

Colour in bodies of water can originate naturally with organic action in the humification process through humic and fulvic acids, or minerally with discharges of industrial waste, iron and manganese compounds. The process of decomposition of vegetation with excessive production of humic acids is one of the contributors to dark rivers (BRASIL, 2014).

2.1.5 Turbidity

It is expressed by the attenuation of a light beam as it passes through a sample, with scattering and absorption of light depending on suspended materials (CETESB, 2015), expressed in turbidity units, Jackson units or nephelometric units.

High turbidity values are generally found in bodies of water that have a high capacity to receive materials by dragging them in the form of clay particles, silt, sand, rock fragments and metal oxides in regions with high rainfall. They can also be influenced by anthropogenic actions such as the discharge of domestic and industrial sewage, and another very important point is the geological characteristics of the drainage basins (BRASIL, 2014).

Results above the specified turbidity level are worrying for organisms, especially those that survive at the bottom of lotic and lentic environments, so they will find it difficult to receive sunlight (CETESB, 2015).

2.1.6 Total solids (TS) and total dissolved solids (TDS)

It refers to the final residue after the process of applying high temperatures over a given time to a water sample (CETESB, 2015). This physical parameter is subdivided into suspended solids plus sedimentable and non-sedimentable solids, which are dispensed with in filtration processes, and dissolved solids, which are able to withstand the filtration process because they are made up of very small particles.

Solids can reach bodies of water naturally through erosive processes, organisms and organic debris, or through anthropogenic actions such as the dumping of domestic and industrial waste (BRASIL, 2014). In high concentrations, they can cause siltation, flooding and damage to aquatic life (CETESB, 2015).

2.2 Chemical quality indicator parameters

2.2.1 Nitrite

Nitrite is a parameter that can be used to determine water quality. Nitrite results indicate the presence of bacteria that use organic matter to produce nitrite. Nitrite is produced by bacteria under favourable conditions, mainly in a state of anaerobiosis, and according to oxygen demand it can be converted into nitrate and regulated according to the thermal behaviour of the water body (GONÇALVES; TOZZO, 2014).

2.2.2 Nitrate

Nitrate is one of the forms in which nitrogen is found in the aquatic environment, mainly in surface waters. Its presence in high quantities can be harmful to human health. It is found in domestic sewage, industrial waste and agriculture, as well as contributing to the contamination of water bodies, with excess nitrogen applied and leached into watercourse beds (MUNIZ; PEREIRA, 2011).

2.2.3 Ammoniacal nitrogen

It is a form of nitrogen present in water bodies and is also known as ionised ammonia. It is due to the process of biological degradation of organic matter, where ammonia is obtained by converting the nitrogen molecule. Its levels occur in low quantities in natural waters, but if high results are found it may indicate pollution through effluents or fertilisers (MUNIZ; PEREIRA, 2011).

2.2.4 Total phosphorus

Phosphorus is one of the most important nutrients for the growth of aquatic plants (CETESB, 2015). It can be found either in soluble organic form through dissolved or particulate matter with microorganism biomass or in soluble inorganic form through phosphorus salts or particulate through mineral compounds such as apatite (BRASIL, 2014). In high concentrations, it induces the growth of photosynthetic aquatic organisms, triggering the eutrophication process (MUNIZ; PEREIRA, 2011).

The presence of phosphorus in water is directly linked to the addition of superphosphate detergents, faecal matter, rainwater drainage from agricultural areas carrying fertilisers into watercourses or through industrial effluents (CETESB, 2015).

2.2.5 Alkalinity

It represents the capacity of water to neutralise acids (ESTEVES, 1998), indicating the buffering capacity of water, i.e. if an aquatic environment receives loads of excessive acidic substances or

With high alkalinity values, this environment will not undergo sudden changes in pH.

The presence of bicarbonates, carbonates and hydroxides are the main constituents of alkalinity, with different pH values for each constitution. High pH results are linked to the respiration rate of microorganisms and the decomposition of organic matter (BRASIL, 2014).

2.2.6 Iron

Iron is an element that is included in the characterisation of the potability of water bodies. Outside the permitted limit, it has no influence on the health of the population; its damage is more directed towards domestic purposes in the appearance of stains on clothes and vases or industrial purposes. High concentrations of iron can also indicate a lack of oxygen, especially in groundwater or at the bottom of ponds (BRASIL, 2014).

2.2.7 Total Hardness

Hardness is defined as the difficulty of water to dissolve soap, due to the action of calcium, magnesium, iron and other substances (ANDRADE, 2005), i.e. the sodium and potassium ions are displaced by the cations present in the water to produce certain substances that adhere to sinks and bathtubs (CROUCH et al, 2004). The concentration of calcium carbonate is equivalent to the total concentration of all the cations, as the calcium and magnesium ions in natural waters are greater than any other metal ion and therefore hardness can be expressed in terms of the concentration of calcium carbonates. Determining hardness is a parameter used to measure water quality, as hard water is harmful for domestic and industrial use (CROUCH et al. 2004).

Hardness is expressed as temporary, permanent and total hardness. Temporary hardness or carbonate hardness is when calcium and magnesium ions combine with bicarbonate ions in the presence of heating, differing from permanent hardness which is expressed as the combination of calcium and magnesium ions with ions of sulphates, chlorides, nitrates and others that give rise to soluble compounds that cannot be removed by heating, the sum of temporary and permanent hardness is classified as total hardness (ANDRADE, 2005). According to its concentration, it can be classified as soft water, medium hard water and very hard water according to Crouch et. al (2004).

CHAPTER 3

OBJECTIVES

3.1 General

- To assess the water quality of the Ouro Creek in the municipality of Caxias/MA, by analysing physical and chemical parameters, in order to determine its classification and suitability for its various uses.

3.2 Specific

- To analyse the quality and changes in the Ouro Creek water ecosystem through physical parameters (temperature, pH, conductivity, true colour, turbidity, total solids and total dissolved solids) and chemical parameters (nitrite, nitrate, ammoniacal nitrogen, total phosphorus, alkalinity, dissolved iron and total hardness);

- Comparing the hydrological quality of the Ouro Creek according to CONAMA Resolution 357/2005 for surface waters.

CHAPTER 4

MATERIAL AND METHODS

4.1 Characterisation of the Study Area

The municipality of Caxias is located in the mesoregion of eastern Maranhão, on the morphostructure associated with the sedimentary soils of the Parnaíba Basin (ARAÚJO, 2012), at an average altitude of 100m and coordinates 4°51'55.90" South latitude and 43°2T19.20" West longitude (GOOGLE EARTH PRO 2016). The municipality has a territorial area of 5,196.771 km^2 (IBGE, 2014) and borders the municipalities of Aldeias Altas, Coelho Neto, São João do Sóter, Codó, Matões, Parnarama and Timon.

The course of the Riacho do Ouro watershed is located on the right side of the Itapecuru River, at coordinates -4°50'35.35" S, -43° 15'9.18" O, (GOOGLE EARTH PRO 2016), with an elevation of 104 metres. Tropical climate prevails in most of the state of Maranhão. It is characterised by an annual rainfall of between 1,600 and 1,800mm. Minimum, average and maximum temperatures are usually high. The annual average is over 24° C (ARAÚJO 2012).

The stream stretches for more than 10 kilometres. It rises in the village of Cabeceira do Ouro and flows into the village of Raiz on the banks of the Itapecuru River (Fig.1). Its water regime is permanent and its watercourse has a fluvial hierarchy of 1ª , 2ª and 3ª order (FILHO; VITTE, 2005).

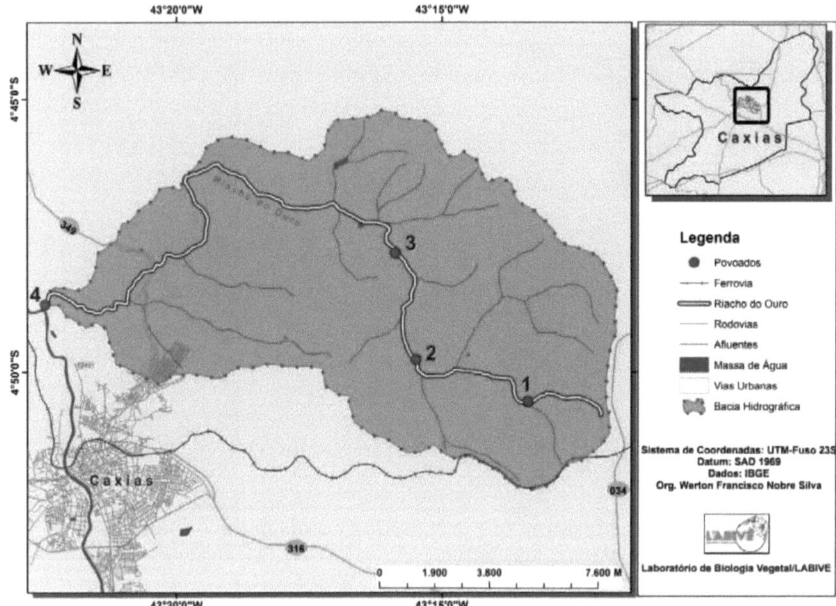

Figure 1 - Location map of the Ouro stream catchment area, Caxias/MA.

(1- Povoado Cabeceira do Ouro, 2- Povoado Bacabalzinho, 3- Povoado Junco, 4- Povoado Raiz)

Source: IBGE, 2006; Organisation: SILVA, W.F.N 2016.

4.2 Definition of sampling points

Four collection points were defined for each site, P1, P2, P3, P4 at the source located in the village of Bacabalzinho (Fig. 2) and in the middle course at points P5, P6, P7, P8 located in the village of Junco (Fig. 3).

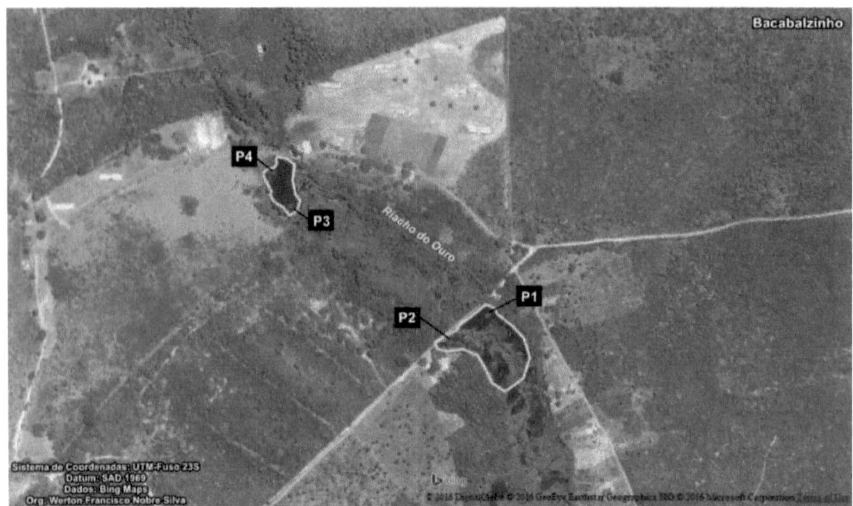

Figure 2 - Map showing the location of the sampling points in the region of the headwaters of the Ouro stream, Caxias/MA.
Source: BING MAPS, 2016; Organisation: SILVA, W.F.N 2016

Figure 3 - Map showing the location of the sampling points in the middle reaches of the Ouro stream, Caxias/MA.
Source: BING MAPS, 2016; Organisation: SILVA, W.F.N 2016

4.3 Collection methodology

On-site expeditions were carried out during the dry season in October and

November 2016 to collect water samples to analyse physico-chemical parameters.

The collection methodology used was in accordance with the manual of collection procedures and methodology for analysing water from Companhia Energética de Minas Gerais (CEMIG, 2009).

The water samples were collected manually using 1.0 litre PET bottles in duplicates, previously identified, approximately 1.0m from the shore at a depth of 20cm. The samples were stored in a Styrofoam box with ice to preserve the physical-chemical characteristics of the water, after which they were sent to the physical-chemical laboratory, except for the air and water temperature parameters, which were carried out on site.

4.4 Analytical methodology for physical indicators

4.4.1 Temperature (°C)

The temperature was measured on site using a calibrated digital thermometer that met the calibration standards. Air and water temperatures were measured at the same collection points.

4.4.2 pH

The pH was determined using a potentiometric method based on the Nernst equation, operated on a digital bench pH meter with a glass reading cell and 0.01 resolution. To guarantee analytical reliability,

Before the analysis, the equipment was adjusted with buffer solutions of pH 4, 7 and 9. Then the test was carried out, where the samples were transferred to 20mL beakers and the electrode was submerged about 6cm into the sample and read.

4.4.3 Conductivity (pS/cm)

It was carried out in accordance with Standard Methods; 19th Edition; 1995; p. Conductivity 2-43, using a benchtop digital conductivity meter with a resolution of 0.1. In the first procedure, the equipment was adjusted with standard solutions of conductivity 147 and 1413μS/cm and then the samples were transferred to 50mL containers and the electrode inserted to take the reading.

4.4.4 True colour (mg Pt/L)

It was determined according to the APHA recommended standard using a spectrophotometric method. The samples were first vacuum filtered through 0.45μ membranes to remove suspended particles (Fig. 4) and then read in a spectrophotometer with a reading range between 190 and 1100nm.

Figura 4. Vacuum filtration system for determining true colour

Source: GASPAR, J.C 2016

4.4.5 Turbidity (UNT)

The turbidity results were obtained using the method based on Standard Methods; 19th ED.; 1995; pp. 2-9 and 2-10, in a 0.01 resolution turbidimeter, with a working range of 0 to 100 EBC (Fig 4). Firstly, the samples were homogenised and transferred to the appropriate glass cuvette and placed in the equipment's reading chamber (a place free from light where the light beams pass through). The results were read in EBC and converted to UNT by calculating NTU = value read in ebc x 4).

Figura 5. Turbidimeter with reading range from 0 to 100 ebc.

Source: GASPAR, J.C 2016

4.4.6 Total solids ST and Total dissolved solids STD (mg/L)

The test was carried out in accordance with Standard Methods; 19th Ed; 1995; p. Solids 2-54 to 2-55. Before the test, the capsules used to pack the samples were preheated in ovens (Fig. 6) for a time interval of 1 hour (104+/-1°C for total solids determination and 180°C for total dissolved solids determination) and cooled in desiccators and weighed, after which the samples were homogenised and 100mL transferred to the capsules and placed back in the oven until they evaporated at a temperature of 98°C, after which they were heated for another hour (104+/-1°C for total solids determination and 180°C for total dissolved solids determination), and finally cooled in vacuum desiccators and weighed. The results were obtained by calculating the masses weighed.

Figure 6: Drying the samples in an oven at the temperatures required for the test.

Source: GASPAR, J.C 2016

4.5 Analytical methodology for chemical indicators

4.5.1 Nitrite (Absence or presence)

The test was carried out qualitatively, i.e. only the absence or presence of nitrite was determined. Two test tubes were used with samples and a blank with distilled water, both of which received the same treatment with the addition of a nitrite indicator solution, where the sample behaved as a blank, thus validating the absence of nitrite.

4.5.2 Nitrate (mg/L)

It was determined using a spectrophotometric method in which the sample and the distilled water blank received the same treatment by adding the Nitra Ver 5 Reagent Power Pillow and after resting the reading was taken on a Hach 2010 spectrophotometer.

4.5.3 Ammoniacal nitrogen (Absence or presence)

Determination of ammoniacal nitrogen was qualitative using the Nessler method, in which the test consisted of a blank with distilled water and a sample, both of which were transferred to test tubes and after the addition of Nessler's reagent was read as absent, as the sample behaved the same as the blank.

4.5.4 Total phosphorus (mg/L)

Total phosphorus was determined using a spectrophotometric method, using specific kits for determining total phosphorus in water (Molybdonate Method with Persulfate Acid Digestion), digested in a COD reactor (Fig. 7).

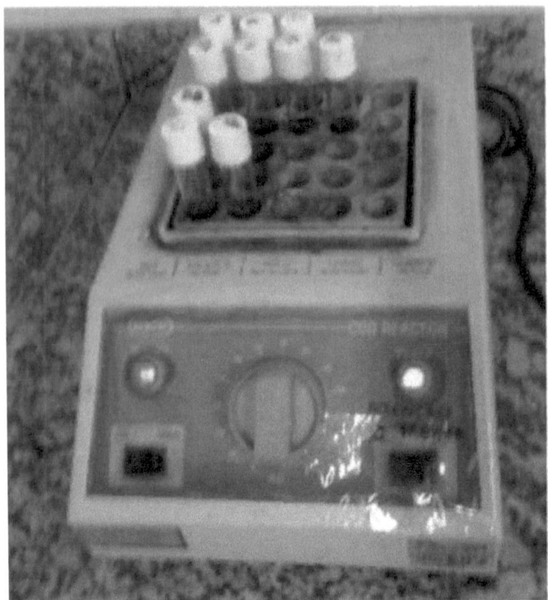

Figura 7. Reactor for sample digestion

Source: GASPAR, J.C 2016

4.5.5 Total alkalinity (mg/L)

It was determined by titration using a pH meter with the electrode immersed in the sample during titration to identify the turning point (Fig. 8).

Figura 8. Titration set with turning point determination with pH meter.

Source: GASPAR, J.C 2016

4.5.6 Iron (mg/L)

The iron content was determined by the spectrophotometric method using Hach's Ferro Mo Iron 1 and 2 Power Pilow reagents and read at a wavelength of 510 nm in a spectrophotometer with a wavelength resolution of 0.1 nm and a reading range of 190-1100nm (Fig. 9).

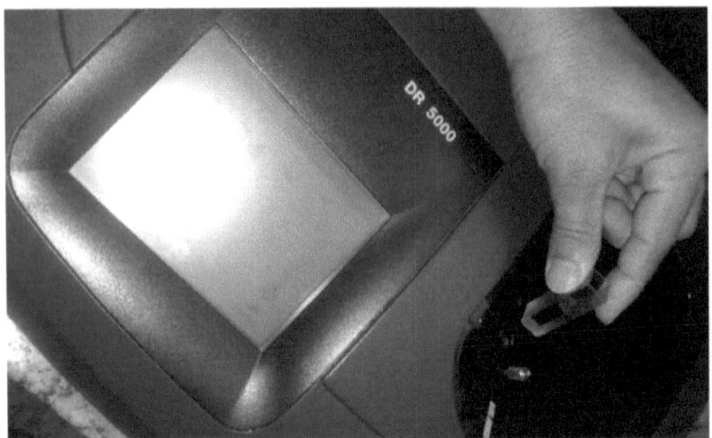

Figure 9. Spectrophotometer with 190-1100nm reading range.

Source: GASPAR, J.C 2016

4.5.7 Total hardness (mg CaCO3/ L)

The determination of total hardness was based on Standard Methods (1995) ; 19th Ed.; 1995; p. Hardness 2-35, by titration with ETDA, sample with 25% alkaline NaOH solution and eriochrome black reagent as an indicator.

CHAPTER 5

RESULTS AND DISCUSSION

5.1 Temperature

Table 1. Air and water temperature (°C) at the source and middle course of the Ouro stream, Caxias, Maranhão.

	Spring				Medium course			
Parameters	**P1**	**P2**	**P3**	**P4**	**P5**	**P6**	**P7**	**P8**
Air Temperature (°C)	30	30	30	30	31	31	31	31
Water Temperature (°C)	26	26	28	28	27	27	27	28

Source: GASPAR, J.C 2016

Table 1 shows that there were no significant variations in air temperature (average 31°C) and water temperature (average 27°C). These minimal temperature fluctuations at the sampling points ensured the accuracy of the results, i.e. the parameters were assessed under the same environmental conditions. These minimal variations in air and water temperature allow bodies of water to balance the biological processes of aquatic life (VASCO et al; 2011; apud DALTRO FILHO; SANTOS, 2001). Temperature is one of the important parameters for physical processes in the solubility of chemical gases in reactions and biological processes in the growth rates of aquatic organisms (ESTEVES, 1998).

However, the average temperature of 27°C can be attributed to the fact that the riparian forest is in the process of being deforested, allowing suspended particles to appear, contributing to the increase in temperature (ESTEVES, 1998; BACELLAR, 2005; KRNO et al., 2013). However, according to Von Sperling (2007), the average temperature of aquatic environments varies between 22 and 30°C in Brazil, which is an important ecological factor that can influence a diversity of aquatic individuals and the similarity that can exist between it and the content of dissolved gases.

5.2 pH

Table 2. pH of the different collection points in the Ouro stream, Caxias, Maranhão and compared with CONAMA resolution 357/2005.

	Spring				Medium course			
Parameters	P1	P2	P3	P4	P5	P6	P7	P8
PH	6,3	6,3	6,2	6,2	7	7,2	6,4	6,4
CONAMA 357/2005	6 a 9							

Source: GASPAR, J.C 2016

Analysing Table 2, it was found that the pH results of the different points compared to CONAMA resolution 357/2005 (BRASIL, 2005) for class 1 waters, were in line with the current standard, with minimum values of 6.2 and maximum values of 7.2. Among these results, it was observed that the values are similar at the different collection points, with the exception of points P5 and P6 which reached a maximum value of 7.2, i.e. within the pH scale, the waters of the Ouro stream are in a neutral to slightly acidic range.

These results of neutral to slightly acidic pH in the Riacho do Ouro are probably influenced by natural actions, although some anthropogenic actions were observed, such as deforestation, burning, silting up, among others, along the entire length of the stream, but none of these can have an influence on sudden changes in pH.

According to Esteves (1998), the aquatic community has a direct influence on pH, citing the assimilation of CO^2 by algae and macrophytes as being part of these processes. It is therefore likely that the influence of the pH values found is due to the large number of aquatic plants in the stream bed, which contributes to the decomposition of organic matter, consequently producing humic acids, as evidenced by the yellowish colour of the water itself (BRASIL, 2014). Another considerable point is the production of carbonic acid by phytoplankton or even the chemical constitution of the soil itself and the dissolution of rocks. Regarding the concentration of carbon dioxide due to the reduction in pH caused by the increase in organic matter, Esteves (2011) explains that this can affect the survival of aquatic organisms, jeopardising the ecological relationships present in that environment.

5.3 Conductivity

Table 3. Conductivity (pS/cm) of the different collection points in the Ouro stream, Caxias, Maranhão.

	Spring				Medium course			
Parameters	**P1**	**P2**	**P3**	**P4**	**P5**	**P6**	**P7**	**P8**
Conductivity (µS/cm)	106	105	643	636	192	200	353	355
CONAMA 357/2005					---------			

*Maximum 500 to 800 reference ANZECC; KPDES, 2010

Source: GASPAR, J.C 2016

Analysing Table 3, it was possible to identify large variations in conductivity between the collection points, with a maximum average result of 639.5|jS/cm for points P3 and P4 and a minimum of 105.5 pS/cm for points P1 and P2 (Fig. 10), all of which are located in the spring. In principle, these abrupt differences in conductivity should not exist in environments with practically the same conditions and the same region. One justification for these variations may be related to various factors such as geological formation, since, according to Arcova; Cicco (1999); Souza; Tundisi (2000); Bassoi; Guazelli (2004), some factors such as geologically poor areas and the absence of point sources of pollution can contribute to low water conductivity. According to Esteves (1998), low pH conditions and high decomposition rates are factors which increase conductivity, but these values are not always high concentrations of ions, but rather ionisable substances which lead to errors in determining conductivity and an increase in the result.

The conductivity results were compared with Anzecc (2010), whose maximum reference limit is 500pS/cm and Kpdes (2010), with 800|jS/cm, because CONAMA resolution 357/2005 does not give relevance to this variable. Esteves (2011), however, points out that conductivity is a prepowering factor that can subsidise the degradation process of a drainage basin.

Figure 10. Collection point P2 located in the source area of the Ouro stream, Caxias, MA.
Source: GASPAR, J.C 2016

5.4 True colour

Table 4. True colour sampling (mg Pt/L) from the different collection points in the Ouro stream, Caxias, Maranhão, compared to CONAMA resolution 357/2005.

	Spring				Medium course			
Parameters	**P1**	**P2**	**P3**	**P4**	**P5**	**P6**	**P7**	**P8**
Colour (mg PV L)	15	16	15	22	61	110	20	20
CONAMA 357/2005					75			

Source: GASPAR, J.C 2016

According to Table 4, point 6 was the only one to show a result higher than that permitted by CONAMA resolution 357/2005 for class 2 freshwater (BRASIL, 2005). This result may be related to the high rate of decomposition of organic matter, roots, sticks and leaves, due to the excessive concentration of aquatic plants triggered by the eutrophication process and consequently the production of humic substances, which is strengthened by the appearance of the yellowish colour of the water, characteristics caused by the presence of this type of substance (UFRJ, 2012).Although the dissolved iron content was high (Table 12), this is not relevant to the increase in colour at points P5 and P6 (Fig. 11) due to the physical characteristics of the water in the Ouro stream, as water rich in iron gives it a purplish appearance with characteristic flavours and odours.

Figure 11. Collection point P5 located in the village of Junco Caxias/ MA

Source: GASPAR, J.C 2016

5.5 Turbidity

Table 5: Turbidity of the different collection points in the Ouro creek, Caxias, Maranhão and comparison with CONAMA resolution 357/2005.

	Spring				Medium course			
Parameters	**P1**	**P2**	**P3**	**P4**	**P5**	**P6**	**P7**	**P8**
Turbidity (UNT)	5,4	1,2	3,4	4,4	7,5	7	30	32
CONAMA 357/2005	40							

Source: GASPAR, J.C 2016

According to table 5, the turbidity results for the Ouro stream met the requirements of CONAMA resolution 357/2005 for class 1 fresh waters (BRASIL, 2005). There was an increase in the values for points P7 and P8 compared to the others, which could probably be influenced by the presence of suspended sandy particles. The Ouro stream has clear, transparent water, as shown by the true colour result, but it is rich in suspended particles, a kind of sandy material with a high grain size and a large amount of mud in the depths. According to Esteves (1998), suspended particles such as phytoplankton and organic and inorganic debris are one of the main characteristics that increase the turbidity result. The turbidity results will not jeopardise aquatic life since the current CONAMA legislation considers this quality parameter to have a maximum limit of 100 UNT.

5.6 Total solids (TS) and total dissolved solids (TDS)

Table 6. Total solids ST and total dissolved solids STD (mg/L) from the different collection points in the Ouro stream, Caxias, Maranhão and compared with CONAMA resolution 357/2005.

	Spring				Medium course				
Parameters	**P1**	**P2**	**P3**	**P4**	**P5**	**P6**	**P7**	**P8**	**357/2005**
ST (mg/L)	50	60	90	80	370	330	320	330	--
STD (mg/L)	10	10	10	10	30	20	20	30	500

Source: GASPAR, J.C 2016

CONAMA resolution 357/2005 only determines specific values for total dissolved solids. In this case, total solids for legislation are irrelevant for determining water quality. With regard to the sampling points mentioned in Table 6, it was found that all the points are within the acceptable limit of current legislation for total dissolved solids. It was also found that there is a greater quantity of particles in the middle course when compared to the points at the source.

The high values for total solids at points P5 and P6 may be related to the devastation of the riparian forest and the small volume of water that can carry organic and inorganic substances into the stream bed (BRASIL, 2014). Points P7 and P8 have a larger volume of water with the presence of plants around them and excess mud on the banks and bottom of the stream, and these factors have a direct effect on the increase in the concentration of dissolved solids.

Table 7. Qualitative results of nitrite sampling from the different collection points in the Ouro creek, Caxias, Maranhão and compared with CONAMA resolution 357/2005.

	Spring				Medium course			
Parameters	**P1**	**P2**	**P3**	**P4**	**P5**	**P6**	**P7**	**P8**
Nitrite (A/P)	A	A	A	A	A	A	A	A
CONAMA 357/2005					1			

*A absence P presence

Source: GASPAR, J.C 2016

Table 7, for the qualitative nitrite test, shows only the presence or absence of nitrite for the different points. All the results obtained in the Ouro stream were absent. The CONAMA 357/2005 resolution for class 1 freshwater has a maximum quantitative acceptable limit of 1.0mg/L (BRASIL, 2005).

Generally, the presence of nitrite in water is low due to the presence of dissolved oxygen, which causes its instability (ESTEVES, 1998). The concentration of nitrite in water bodies can also be caused by bacteria in anaerobic conditions which oxidise nitrate due to the availability of oxygen in the water, according to Neto (2003). The fact that there is no evidence of pollution from sewage systems and the instability of this ion may explain its absence throughout the study area. The study found that the nitrite result for the Ouro stream is within the acceptable limit of CONAMA 357/2005.

Table 8. Nitrate (mg/L) from the different collection points in the Ouro stream, Caxias, Maranhão and compared with CONAMA resolution 357/2005.

	Spring				Medium course			
Parameters	**P1**	**P2**	**P3**	**P4**	**P5**	**P6**	**P7**	**P8**
Nitrate (mg/L)	0	0	0	0	0,8	0,6	0,1	0,1
CONAMA 357/2005	10							

Source: GASPAR, J.C 2016

The nitrate results at the different sampling points complied with CONAMA 357/2005, which sets a maximum limit of 10 mg/L. According to Table 8, P5, P6, P7 and P8 (Fig. 12) had nitrate levels at the sampling points. The presence of nitrate may be related to the marginal use of land for agriculture on the banks of the stream, with the use of fertilisers. According to Baird (2002), intensive cultivation of the land facilitates the oxidation to nitrate of the reduced nitrogen present in decomposed organic matter due to the effect of aeration and humidity.

Figure 12. Collection site located at points P7 and P8 in the Junco locality.

Source: GASPAR, J.C 2016

Table 9. Qualitative results of ammoniacal nitrogen sampling from the different collection points in the Ouro creek, Caxias, Maranhão and compared with CONAMA resolution 357/2005.

	Spring				Medium course			
Parameters	**P1**	**P2**	**P3**	**P4**	**P5**	**P6**	**P7**	**P8**
Ammonia (A/P)	**A**	A	A	A	A	A	A	A
CONAMA	3,7							

357/2005

*A absence P presence

Source: GASPAR, J.C 2016

According to Table 9, the amount of ammoniacal nitrogen was absent for all the points. In this way, the results are in line with current CONAMA 357/2005 legislation for class 1 fresh waters with a pH < 7.5, which has a maximum limit of 3.7mg/L (BRASIL, 2005).

The amount of ammoniacal nitrogen is lower than that of nitrate (ESTEVES, 1998), where results on average were higher than those for ammoniacal nitrogen. According to Vasco (2011), ammoniacal nitrogen is an indicator of organic pollution from domestic waste, which may justify the results obtained as absent for these ions. In this context, the communities living around the Ouro stream do not dump domestic waste, a fact that was observed during the collection period, thus demonstrating the level of environmental awareness of the residents. According to Santos et al. (2001), the conservation of natural resources must take place both for ecological reasons and for socio-economic reasons, since society maintains a degree of dependence on it for its own personal well-being and in general for everyone.

5.7 Total phosphorus

Table 10. Total phosphorus (mg/L) from the different collection points in the Ouro stream, Caxias, Maranhão and compared with CONAMA resolution 357/2005.

	Spring				Medium course			
Parameters	**P1**	**P2**	**P3**	**P4**	**P5**	**P6**	**P7**	**P8**
Phosphorus (mg/L)	0	0	0	0	1,6	2,4	2,7	2,3
CONAMA 357/2005	0,1							

Source: GASPAR, J.C 2016

The results for total phosphorus in the Ouro stream were different for the headwaters and the middle course, with the results for the headwaters being within the norms, while for the middle course they were outside the limits defined by CONAMA resolution 357/2005 for class 1 fresh water. The geomorphology of the points analysed is basically the same, and this reduces the likelihood of the increase in total phosphorus concentration being caused by natural causes, where they are more directed towards anthropogenic actions. According to Table 10, points P5, P6, P7 and P8 showed agro-industrial and agricultural activities, such as the cultivation of sugar cane for the production of artisanal cachaça, grazing and pig farming. These activities can generate substances from the fertilisers used on pastures and sugar cane plantations and faecal waste from pig farming, all of which can reach the water table or be carried to the stream bed and contribute positively to the increase in total phosphorus (ESTEVES, 1998).

One of the reasons for the increase in phosphorus at points P5, P6, P7 and P8 could be the process of eutrophication. According to Esteves (1998), phosphorus serves as a nutrient and contributes positively to increased plant development, thus contributing to this process. At these points, there was a high density of plants in the bed and on the banks of the stream and a yellowish colour resulting from decomposition processes. Large populations of organisms can affect aquatic life by increasing oxygen consumption and increasing the number of anaerobic organisms.

According to Rodrigues (2007), anthropogenic actions can compromise water quality, causing alterations such as a reduction in dissolved oxygen, an increase in turbidity, changes in pH and an increase in the amount of nitrogen and phosphorus. From this perspective, it can be inferred that the alterations found at these points are affecting the quality of the water, contributing to the increased environmental degradation of the stream and jeopardising the survival of aquatic organisms.

5.8 Total alkalinity

Table 11. Alkalinity (mg/L) of the different collection points in the Ouro stream, Caxias, Maranhão.

	Spring				Medium course			
Parameter	**P1**	**P2**	**P3**	**P4**	**P5**	**P6**	**P7**	**P8**
Alkalinity	10	10	10	15	40	48	13	12,4
CONAMA 357/2005					-----			

*Reference 30 to 500. Source: COELHO, et al.2015, apud MORAES, 2000 **Source:**

GASPAR, J.C 2016

Table 11 shows the total alkalinity results for the waters of the Ouro stream, where all the points were found to be low. The National Environmental Council CONAMA 357/2005 does not establish total alkalinity as an indicator parameter for surface water quality. However, Coelho, et. al (2015) apud Moraes (2008), cite the importance of the processes of decomposition of organic matter and respiration of microorganisms values of 30 to 500mg/ L.

The pH values increase the accuracy of the total alkalinity results by increasing the results for both parameters at the same points (P5 and P6), i.e. the higher the pH value, the higher the total alkalinity result. According to Table 2, the pH values found were between 6.2 and 7.2, according to BRASIL (2014) pH results between 4.4 and 8.3 represent the presence of bicarbonates only. According to Coelho, et. al (2015) apud Moraes (2008), only points P5 and P6 were within acceptable limits.

5.9 Iron

Table 12. Dissolved iron (mg/L) from the different collection points in the Ouro stream, Caxias, Maranhão and compared with CONAMA resolution 357/2005.

	Spring				Medium course			
Parameters	**P1**	**P2**	**P3**	**P4**	**P5**	**P6**	**P7**	**P8**
Iron (mg/L	0,3	0,38	0,3	0,36	1,9	1,8	0,7	0,72
CONAMA 357/2005					0,3			

Source: GASPAR, J.C 2016

Table 12 shows the dissolved iron content for the different sampling points in the Ouro stream and compares them with CONAMA resolution 357/2005 for class 1 fresh water. Among the points analysed, it was found that only two points (P1 and P3) were within the acceptable limits of current legislation. The other points obtained results outside the acceptance limit. The variations in results between the sampling points may have been influenced by the volume of water, since the smaller the volume, the higher the concentration.

The amount of iron in the water may originate from the constitution of the soil type itself, the yellow latosol type is the predominant soil in the study area (IBGE, 2015), which contains a significant amount of iron that can be dissolved in the water through the weathering process.

Although the iron content in the Ouro stream does not comply with current legislation, these values do not harm the health of the population using the water resource or aquatic populations (BRASIL, 2014).

5.10 Total hardness

Table 13. Total hardness (mg/L) of the different collection points in the Ouro stream, Caxias, Maranhão and compared with Ministry of Health resolution 518/2004.

	Spring				Medium course			
Parameters	**P1**	**P2**	**P3**	**P4**	**P5**	**P6**	**P7**	**P8**
Hardness (mg/L)	28	31	30	30	8,6	10	17,5	18
CONAMA 357/2005				-----				

*Reference: maximum 500 (Ministry of Health Ordinance). Source: GASPAR, J.C 2016

Source: GASPAR, J.C 2016

The total hardness parameter is not contained in CONAMA resolution 357/2005. However, the data was compared with Ministry of Health regulation 518/2004 for water intended for human consumption, with a maximum acceptable limit of 500 mg/L.

According to Table 13, the total hardness results obtained for the Ouro stream are within the acceptable limit of the Ministry of Health's legislation. Waters with a concentration of < 50 mg/L of $CaCO_3$ are classified as soft (BRASIL, 2014). The Ouro stream has an average of 22 mg/L for the total hardness variable. In general, the results found for the parameters assessed are within the acceptance limits (Fig.13). These results are significant for determining the quality of the Ouro stream.

Figure 13. Quality indicators for each point in the study area.

Indicators	P1	P2	P3	P4	P5	P6	P7	P8
pH								
Conductivity								
True colour								
Turbidity								
STD								
Nitrite								
Nitrate								
Ammoniacal nitrogen								
Total phosphorus								
Alkalinity								
Iron								
Total hardness								

Meets specifications

Does not meet specifications

Source: GASPAR, J.C 2016

CHAPTER 6

CONCLUSION

It was possible to observe the quality of the water in the Ouro stream, as well as the physical and chemical concentrations of the substances and compare them with CONAMA Resolution 357/2005.

The concentration of total phosphorus was outside the acceptance limit at the points in the middle course, probably as a result of anthropogenic inputs and actions from agricultural activities.

For total alkalinity, the results were lower than the acceptance limit, which may be due to the natural processes of decomposition rates and consequently susceptible to pH changes, depending on the amount of acidic or basic material received, while the iron content may be associated with the constitutions of the predominant soil type in the region.

Nitrogen in its different forms, nitrite, nitrate and ammoniacal nitrogen, showed results that comply with the current resolution of the National Environment Council, indicating that the Ouro stream is in minimal or absent conditions for some of the indicator substances for this type of pollution from domestic activities, effluents and agriculture.

Point P6 obtained a value above that permitted by law for the true colour parameter, pointing to natural actions influenced by the amount of decomposing material in the stream bed associated with the low volume of water. According to the total hardness parameter, the Ouro stream is classified as soft water.

The evaluation of the physico-chemical parameters of the water quality of the Ouro creek serves to alert the public authorities to take decisions with a view to conservation campaigns aimed at Caxiense society, especially the communities that surround it, with a view to using it sustainably in order to protect the biodiversity and biological balance of aquatic organisms and those that depend on this water resource for their survival.

REFERENCES

ALMEIDA, A. C. P. P.; CARVALHO, M. D.; RAMOS, S. M.; MOTA, H. R. **Manual of collection procedures and water analysis methodologies: SISAGUA**. Belo Horizonte: CEMIG, 2009.

ANDRADE. C. A. F. **Analysis of the quality of groundwater used for irrigation in the municipality of Baraúna-RN.** Master's dissertation. Mossoró, 2005.

ANZECC- AUSTRALIAN AND NEW ZEALAND ENVIRONMENT AND CONSERVATION COUNCIL. **Australian Water Quality Guideles for Fresh and Marine Waters, National Water Quality Management Strategy.** ANZECC, Canberra, 2010.

APHA- American Public Heath Association. **Standard methods for the examination of water and wastewater.** Washington. 1995.

ARCOVA, C. S.; CICCO, V. Water Quality of Microbasins with Different Land Uses in the Region of Cunha, State of São Paulo. **Scientia forestalis**, v. 56, p. 125-134, 1999.

BACELLAR L.A.P. **O papel das florestas no regime hidrológico de bacias Hidrográficas.** Geo. br, Ouro Preto, v. 1, p. 1-39, 2005. http//:www.degeo.ufop.br/geobr.

BAIRD,C. **Environmental Chemistry**.2002. 2ª ed., Porto Alegre, Bookman,622p.

BARBOSA, R. M. P.; BOMFIM, O. E.; FERNANDES, L. A.; JUNIOR, R. H. A.; GADELHA, C. L. M.; SILVA, E. S. **Analysis of the variation in BOD, COD and DO in the Mussuré/PB stream.** NORTHEAST WATER RESOURCES SYMPOSIUM. João Pessoa, 2012.

BARROS, V. L. L.; SILVA, P. J.; PEREIRA, L. C.; VALE, F. S. Rio Itapecuru: Uma visão geoambiental em Caxias-MA. **Humana Magazine.** Paço do Lumiar, v. 1, n, 2, p. 104-119, 2014.

BASSOI, L. J.; GUAZELLI, M. R. **Environmental control of water.** In: PHILIPPI, JR., A., ROMÉRIO, M. A.; BRUNA, G. C. Curso de gestão ambiental. Monole. Barueri, 2004. 1045p.

BRAZIL. Law No. 9.433 of 8 January 1997. **Institutes the National Water Resources Policy and creates the National Water Resources Management System.** Brasília/DF, Brazil. Ministry of the Environment. 1997.

BROWN, J. H.; LOMOLINO, M. V. **Biogeografia**. 2. ed. Editora Funpec. Ribeirão Preto, 2006.

BRUNI, J. C. Water and life. **Rev. Social,** USP, São Paulo, 1993.

Cities. Available at < http.www.ibge.gov.br.> Accessed on 01.12.2016.

COELHO.D.A, SILVA.A.R.S, CASTRO.T.O, SANTOS.R.C.G, PASSOS.A.S. **Analysis of total alkalinity and inorganic carbon concentration in urban river stretches: The example of the Santa Rita River, south-western region of Bahia.** VI Brazilian Congress on Environmental Management, Porto Alegre, 2015.

ENVIRONMENTAL COMPANY OF THE STATE OF SÃO PAULO-CETESB. IQA online calculation. 2015. Available at: <http://sobreasaguas.info/iqa_cetesb.aspx>. Accessed on: 13 October 2016.

NATIONAL ENVIRONMENTAL COUNCIL (CONAMA). Resolution no. 357 of 17 March 2005. **Ministry of the Environment**. 2005.

COSTA.S.A.D. **Chemical, physical and mineralogical characterisation and classification of iron-rich soils in the Iron Quadrangle.** Postgraduate project in soils and plant nutrition. Federal University of Viçosa. Viçosa, 2003.

CROUCH, HOLLER, SKOOG, WEST. **Fundamentals of analytical chemistry.** 8. ed. North American translation. 2004.

DETONI, T. L.; DONODONI, P. C.; PADILHA, E. A. Water scarcity: A global look at sustainability and academic awareness. **XXVII National Meeting of Production Engineering**. Foz do Iguaçu-PR, 2007.

ESTEVES, F. A. Fundamentals of Limnology. **Interciência**. Rio de Janeiro, 2011. p. 826.

ESTEVES. F.A. **Fundamentos de limnologia. 2ª** Ed. Rio de Janeiro: Interciência, 1998.

NATIONAL HEALTH FOUNDATION. **FUNASA**. Practical manual for analysing water. 4. ed. Brasília: **FUNASA**, 2013.

GONÇALVES, E. A.; TOZZO, R. A. **Physico-chemical analysis of the water in the Japira stream located in the municipality of Apucarana-PR.** Environment and Sustainability Magazine. Apucarana-PR, 2014.

GUEDES, H. A. S. et al. **Application of multivariate statistical analysis in the study of the water quality of the Pomba River, MG.** Brazilian Journal of Agricultural and Environmental Engineering, p. 558-563, 2012.

KPDES-KENTUCKY POLLUTANT DISCHARGE ELIMINATION SYSTEM. 2010. **Conductivity and Water Quality**. http://kywater.org/ramp/rmcond.htm.

KRNO, I.; SPORKA, F.; STEFKOVA, E. **The influence of environmental variables on larval growth of stoneflies (Plecoptera) in natural and deforested streams.** Biologia, v. 68, n. 5, p. 950-960, 2013.

MARANHÃO. Law No. 8.149/04 of 15 June 2004. **Provides for the State Water Resources Policy, the Integrated Water Resources Management System, and other measures.** São Luís, 2004.

MINISTRY OF HEALTH. National Health Foundation. **Water quality control manual for technicians working in wastewater treatment plants** / Ministry of Health, National Health Foundation-Brasília: Funasa, 2014.

MINISTRY OF HEALTH. **Surveillance and control of the quality of water for human consumption.** Health Surveillance Secretariat. Brasília. 2006 P.212.

MINISTRY OF PLANNING, BUDGET AND MANAGEMENT. **Technical manual of pedology.** 3.ed. Brazilian Institute of Geography and Statistics. Rio de Janeiro, 2015.

MUNIZ, D. H. F.; PEREIRA, M. C.; PARRON, L. M. **Manual of sampling procedures and physical-chemical water analysis.** Documents 232. 1st ed. Colombo PR. EMBRAPA Florestas, 2011.

NETO, M.S.S et al. **Hydrogeochemical characterisation of the Manso-Cuiabá river basin, MG. Acta limnológica Brasiliensia**. vol. 14, p. 14-36, 2003.

ROCHA, A. H.; ROCHA, C. J.; **Introduction to environmental chemistry.** 2 ed. Porto Alegre, 2009.

RODRIGUES, E. **Século XXI rumo ao desenvolvimento sustentável e a educação ambiental.** Revista Bioética e Educação, Belo Horizonte: Bio Consulte, n°. 01,2007, p. 179-190 (Federal University of Lavras - UFLA).

SÁNCHEZ, L. E. **Environmental impact assessment**: concepts and methods. São Paulo: Oficina de Textos, 2008.

SANTOS, J. E.; NOGUEIRA, F.; PIRES, J. S. R.; OBARA, A. T.; PIRES, A. M. Z. C. R. **The value of the ecological station of Jatai's ecosystem services and natural capital.** Revista Brasileira de Biologia, v. 61, n. 2, p. 171-190, 2001.

SECRETARY OF STATE FOR THE ENVIRONMENT AND NATURAL RESOURCES. **Action plan for the prevention and control of deforestation and fires in the state of Maranhão.** Decree no. 27. 317, 14 April 2011, São Luís, 2011.

SOUZA, A. D. G.; TUNDISI, J. G. **Hidrogeochemical comparative study of the Jaú and Jacaré-Guaçu River Watersheds São Paulo, Brazil.** Revista Brasileira de Biologia, v.60, n. 4, p. 563-570, 2000.

UFRJ. **Colour of water**. Rio de Janeiro, 2012. Available at: http//www.ufrrj.br/institutos/it/de/acidentes/cor.htm.

VASCO, A.N.; BRITTO, F.B.; PEREIRA, A.P.S.; MÉLL JÚNIOR, A.V.; GARCIA, C. A.; NOGUEIRA, L.C. **Spatial and temporal assessment of water quality in the Poxim River sub-basin, Sergipe, Brazil**. Ambi-Agua, Taubaté, v. 6, n.1, p.118-130,2011.

VICTORINO, C. J. A. **Planeta água morrerendo de sede**: uma visão analítica na metodologia do uso e abuso dos recursos hídricos. Porto Alegre: EDIPUCRS, 213p. 2007.

VON SPERLING, M. **Studies and modelling of river water quality**. Principles of Biological Wastewater Treatment. 3.ed. Belo Horizonte: UFMG, v. 7, p. 588, 2007.

WALDMAN, M. **Triad of water, waste and energy:** Essentiality of water resources. Environmental Training Course, São Bernardo do Campo-SP, 2009.

Printed by Books on Demand GmbH, Norderstedt / Germany